FUN WITH
NATURE

Written by:	Dympna Hayes Melanie Lehmann
Editor:	Teri Kelly
Design & Illustration:	Annelies Davis

Copyright © 1987 by Hayes Publishing Ltd.

ISBN 0-88625-154-0

CHP BOOKS

3312 Mainway, Burlington, Ontario L7M 1A7, Canada
2045 Niagara Falls Blvd., Unit 14, Niagara Falls, NY 14304, U.S.A.

N

ature is for you to explore and enjoy, so get down on your hands and knees. What can you see?

L ook at the trees.
Notice the leaves, bugs
and birds. Try to spot three
different types of birds. A library
book about birds will help you
with their names. What colors
are the birds?

After a rain shower go hunting for worms. Worms breathe through their skin, so they have to come to the surface when it rains. If they stayed underground in the wet, soggy soil, they wouldn't be able to breathe. Can you think of other changes that happen in nature because of rain?

Lift a stone and see what lives underneath it. Wait and watch, but don't forget to put the rock back after you have had a look!

See if you can find all these things to collect:

- 4 different leaves
- 1 small twig
- 4 pebbles
- 1 blade of grass the length of your hand
- 2 different flowers
- 1 root from a weed

S it quietly and listen.

What sounds do you hear?
Try to identify who or what
is making each sound.

Make a small hole in a spider's web and watch what happens.

D rop a crumb of food in an ant's path. Watch the ant carry the food home. What happens when you run your finger across the ant's path and the ant reaches your scent?

G o exploring and bring some bugs home for a short visit.
Grasshoppers, ladybugs, caterpillars and frogs all make interesting
house guests! Watch for awhile and make sure you set your tiny friends free.

WATER

L̲ook at the sky. Use your imagination. What forms can you see in the clouds? You might see a horse, cat, train, or maybe even a boat!

To play shadow tag, one person is chosen as the shadow chaser and all the other players are runners. When the chaser steps on a runner's shadow, a tag is made. The tagged player becomes the shadow chaser, and the first chaser then becomes the runner.

Using a small butterknife, make a slit about halfway through the stem of one flower. Pull another flower between the slits.

To continue the chain, make a similar split in the second flower and slip a third flower through. Continue until the chain is the length you wish. Attach the last stem to the first with a paper clip. Now your flower crown is ready to wear!

See how many different types of shells, flowers, pebbles, feathers or leaves you can find.

Empty egg cartons make good organizers for small shells and pebbles, and as your collection grows, you can use more egg cartons to store your treasures.

14

Collect a few leaves with interesting shapes and colors. Leaves and flowers usually shrivel up when they dry, so place them between the pages of a thick book for a few weeks and they'll dry flat.

I sn't it nice to sit under a big tree and read a book in the shade? If you sit very still, a few nature friends may join you.

Ask an adult to help you make your own tree gym. A tire swing or rope ladder are great fun and super exercise!

Rake the leaves and make a huge pile. Take a running leap, and whoosh! Scramble in the leaves! Don't forget to rake them up afterward!

The fall is harvest time. Help pick apples, and don't forget to choose your pumpkin for a jack-o-lantern!

P ut some food out for hungry winter animals. Sunflower seeds make good food for tiny creatures. Watch from your window to see who visits.

Build a snowman and make an angel outline in the snow beside it. Lie down on your back in the snow and move your arms and legs up and down. Stand up, and there's your snow angel!

Have fun splashing in a puddle. Bring your boats and ducks along with you. Don't forget to wear your rain boots!

Explore your neighborhood park. There are lots of swings and slides for you to play on.

Pick a beautiful bouquet of flowers. How do they smell? What colors are they?

L

ook for all the new baby animals of springtime
– but don't touch!

C

atch a butterfly. Look at the pretty colors and then set it free.

With help from your parents, plant your own garden with all your favorite vegetables and flowers.

B uild your own sandcastle at the beach. Decorate it with all the pretty seashells and pebbles you can find.

N ature provides many hiding places. Get all your friends together for a game of hide and seek.

Fly a kite to catch the wind! See how high you can make your kite soar!

Set up a tent in your backyard and camp out. Bring some yummy snacks along, and don't forget your flashlight!

Aren't nature's gifts beautiful?

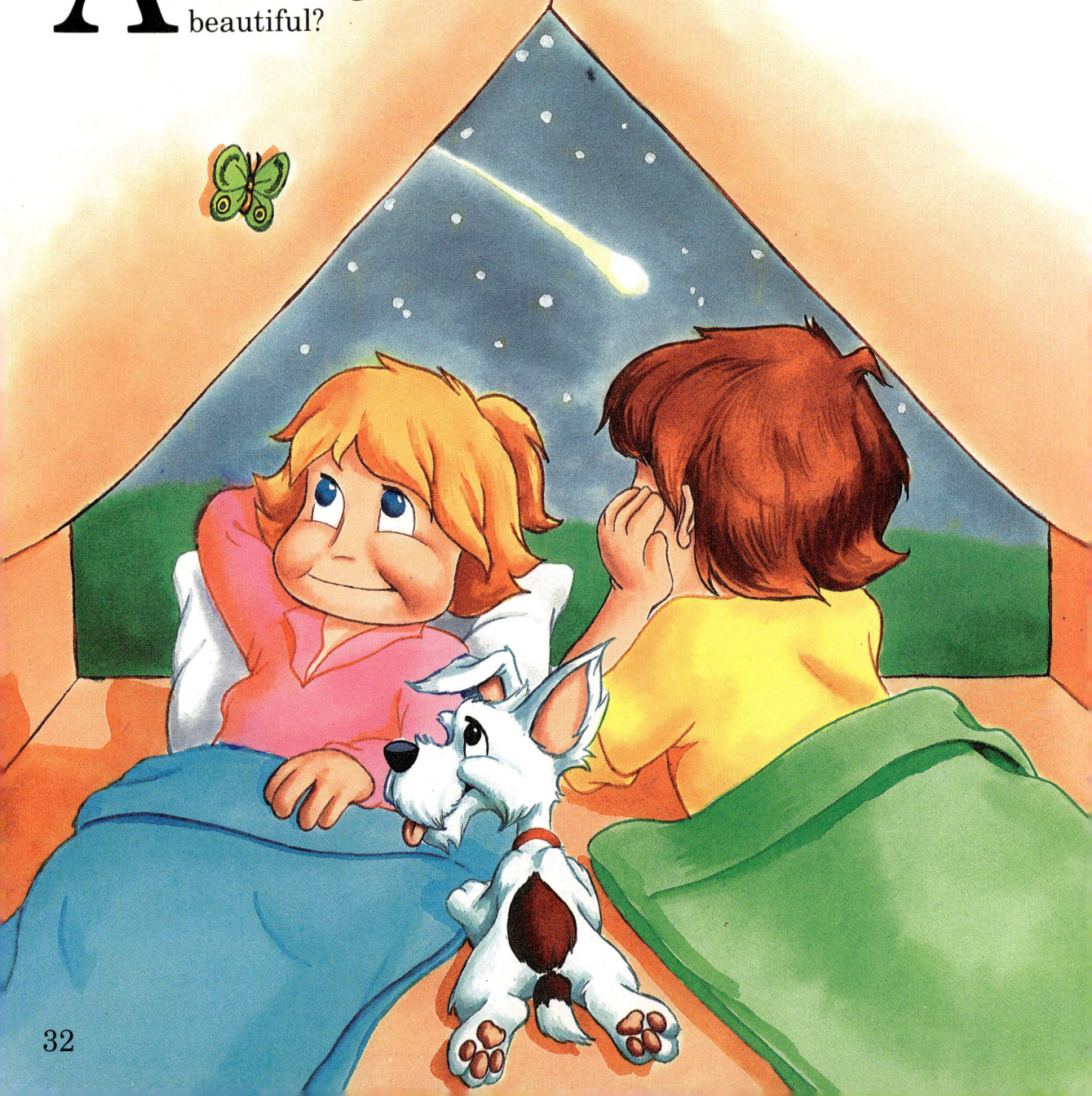